CORAL REEFS

BY

GAIL GIBBONS

New and Updated

HOLIDAY HOUSE ・ NEW YORK

To Mary Cash

Special thanks to Dr. Kim Ritchie, marine biologist;
Nadine Slimak, public relations manager at the
Mote Marine Laboratory, Sarasota, Florida;
and Dr. Amy Apprill, Woods Hole Oceanographic Institute.

The samples of sea life shown in this book are from coral reefs
around the world. These plants and animals do not necessarily
share the same habitat simultaneously as illustrated here. All the
sea life shown in the Great Barrier Reef (pages 26–27) actually
live there.

The Library of Congress has cataloged the previous edition as follows:

Library of Congress Cataloging-in-Publication Data
Gibbons, Gail.
Coral reefs/ by Gail Gibbons. 1st ed.
p. cm .
ISBN 978-0-8234-2080-3 (hardcover)
ISBN 978-0-8234-2278-4-(paperback)
1. Coral reef ecology—Juvenile literature.
2. Coral reefs and islands—Juvenile literature. I. Title.
QH541.5.C7G53 2007
577. 7'89—dc22
2006037959

ISBN 978-0-8234-4370-3 (new and updated hardcover)
ISBN 978-0-8234-4357-4 (new and updated paperback)

Sunlight shines through seawater into a coral reef. The reef is an underwater world of brilliant colors and strange shapes.

Where Coral Reefs Are Found

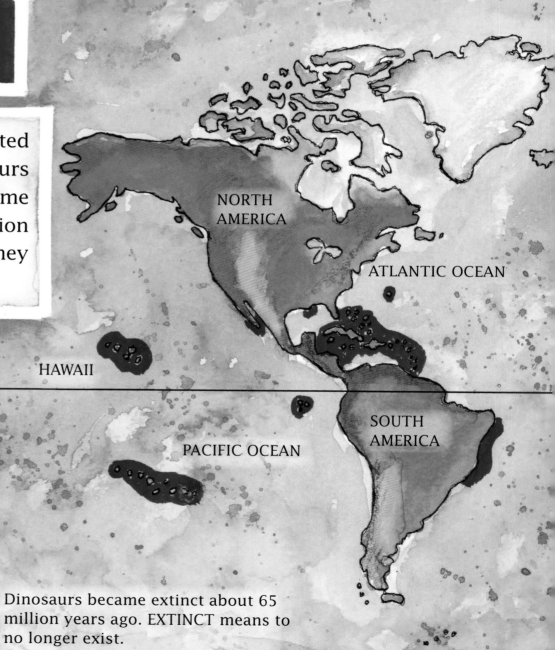

MARINE BIOLOGISTS study the oceans and what lives in them.

Marine biologists believe coral reefs existed about 400 million years ago, when dinosaurs lived. At the same time that dinosaurs became extinct, coral reefs died out. About 50 million years ago coral reefs began to return, and they continue to survive.

NORTH AMERICA

ATLANTIC OCEAN

HAWAII

SOUTH AMERICA

PACIFIC OCEAN

Dinosaurs became extinct about 65 million years ago. EXTINCT means to no longer exist.

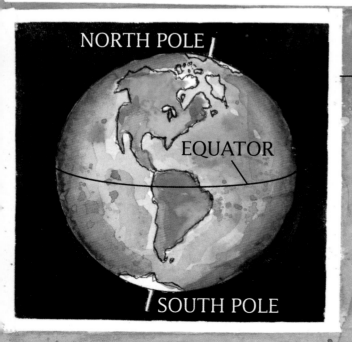

NORTH POLE

EQUATOR

SOUTH POLE

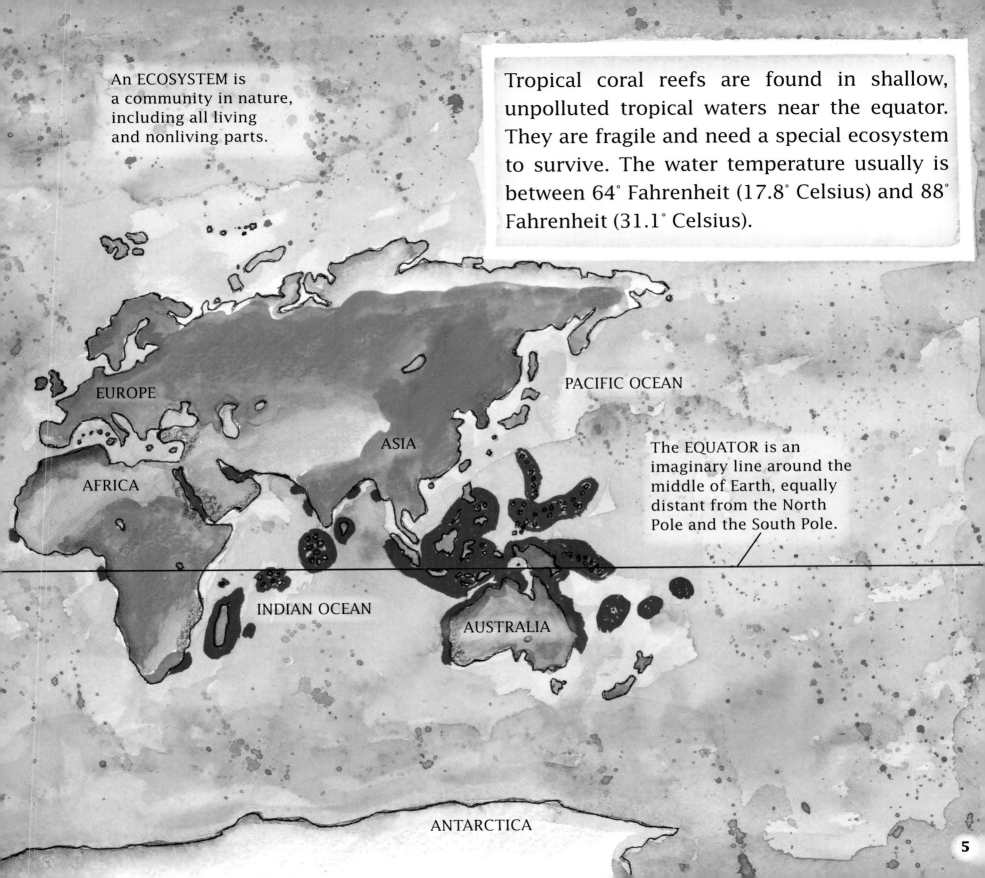

An ECOSYSTEM is a community in nature, including all living and nonliving parts.

Tropical coral reefs are found in shallow, unpolluted tropical waters near the equator. They are fragile and need a special ecosystem to survive. The water temperature usually is between 64° Fahrenheit (17.8° Celsius) and 88° Fahrenheit (31.1° Celsius).

EUROPE

PACIFIC OCEAN

ASIA

The EQUATOR is an imaginary line around the middle of Earth, equally distant from the North Pole and the South Pole.

AFRICA

INDIAN OCEAN

AUSTRALIA

ANTARCTICA

Marine biologists tell us that coral reefs were built over long periods of time with the tiny skeletons of animals called coral polyps.

DIADEM DOTTYBACKS

LONGSNOUT SEA HORSE

LEAFY CORAL

ANGLERFISH

ELKHORN CORAL

FINGER CORAL

Coral Skeleton

Coral polyps produce hard skeletons. When the polyp dies, the skeleton remains, leaving behind the limestone structure that forms habitat for other reef life.

CUP CORAL

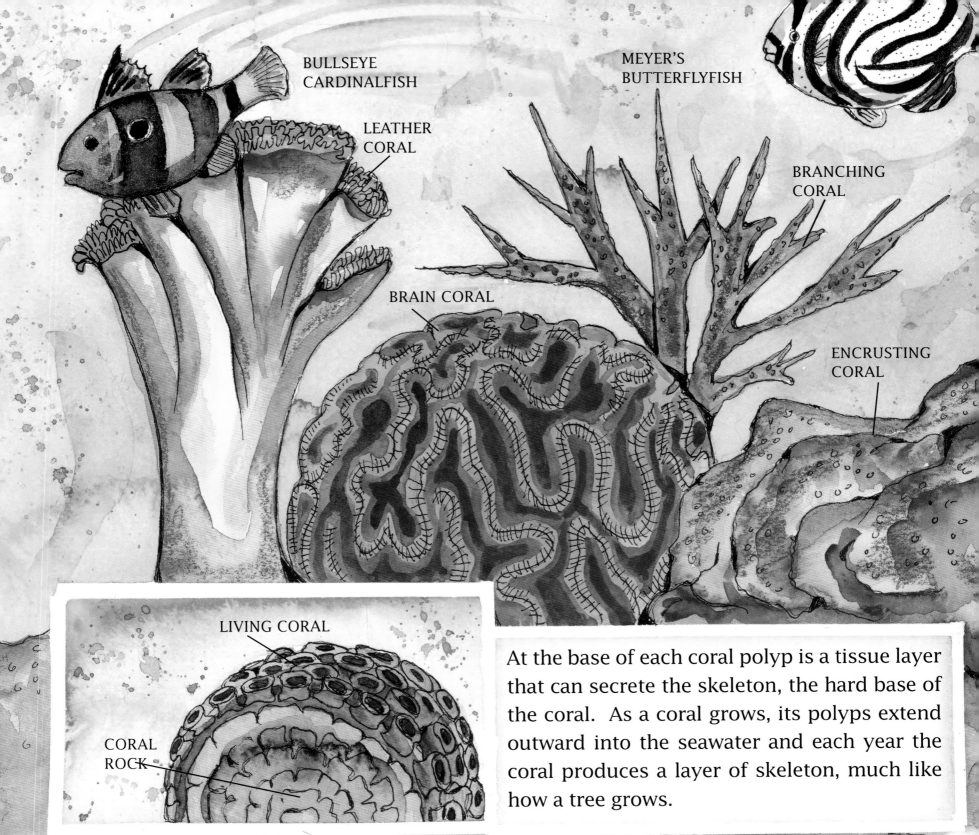

BULLSEYE
CARDINALFISH

MEYER'S
BUTTERFLYFISH

LEATHER
CORAL

BRANCHING
CORAL

BRAIN CORAL

ENCRUSTING
CORAL

LIVING CORAL

CORAL
ROCK

At the base of each coral polyp is a tissue layer that can secrete the skeleton, the hard base of the coral. As a coral grows, its polyps extend outward into the seawater and each year the coral produces a layer of skeleton, much like how a tree grows.

Three Kinds of Coral Reefs

BLUE
DEVILFISH

A fringe reef grows close to the shoreline. A barrier reef grows farther away from the shoreline. It is separated from the shoreline by a channel.

Fringe Reef

Barrier Reef

SHORELINE

Fringe Reef

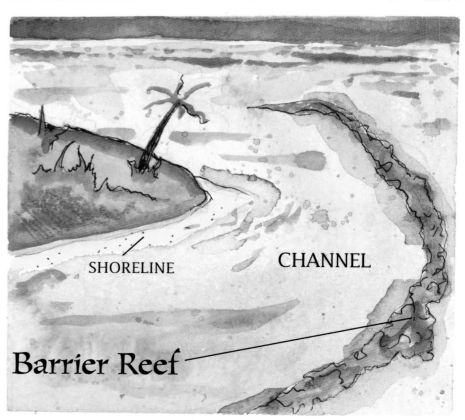

SHORELINE

CHANNEL

Barrier Reef

Atoll

Atoll

An atoll is a reef that is usually formed on the rim of an extinct volcano.

SHORELINE

LAGOON

Tiny pieces of dead corals become SAND.

An atoll is a ring or part of a ring of coral surrounding a lagoon.

Coral Reef Zones

Different species of corals and sea life thrive in their own special places in a coral reef, called zones. Every reef can be divided into three zones.

Shore Zone

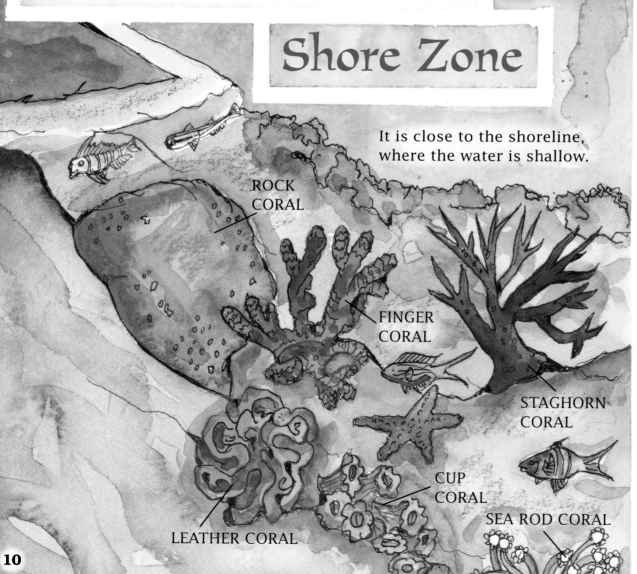

It is close to the shoreline, where the water is shallow.

ROCK CORAL

FINGER CORAL

STAGHORN CORAL

CUP CORAL

SEA ROD CORAL

LEATHER CORAL

Crest Reef Zone

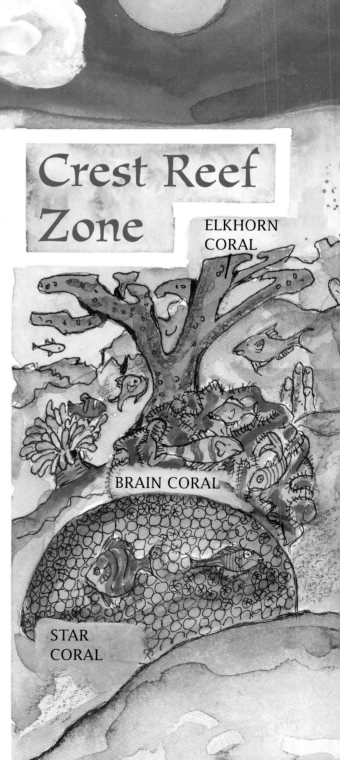

ELKHORN CORAL

BRAIN CORAL

STAR CORAL

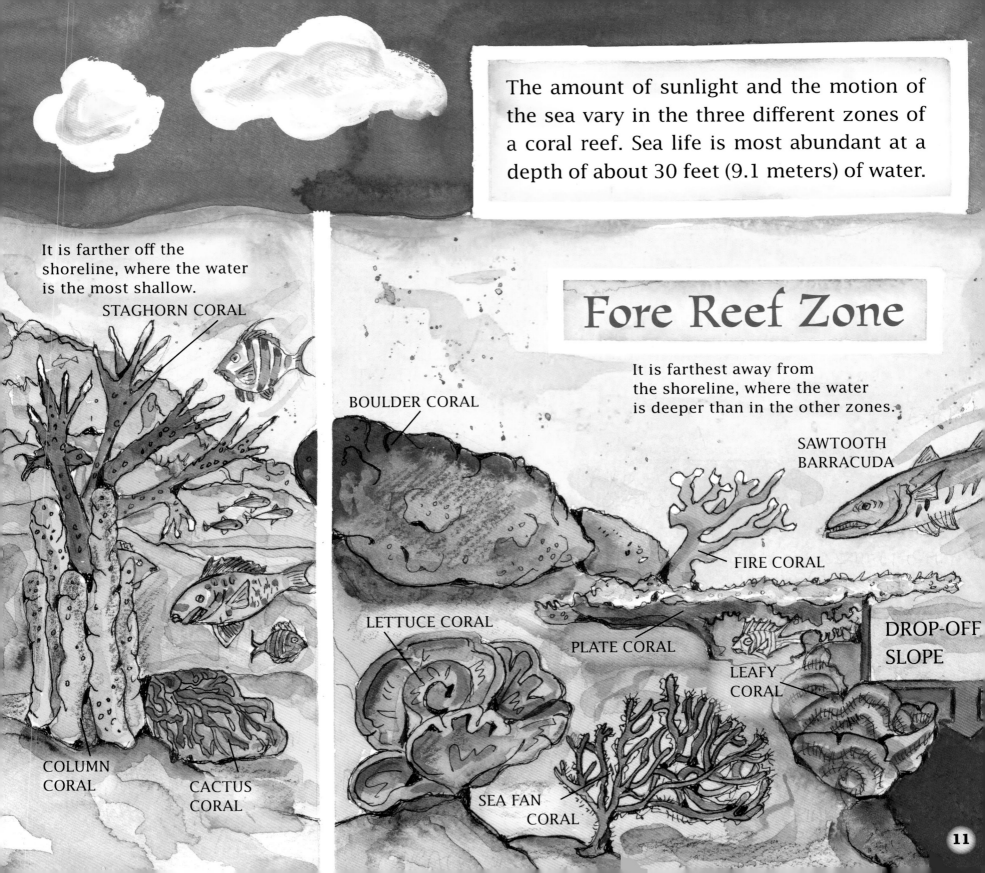

The amount of sunlight and the motion of the sea vary in the three different zones of a coral reef. Sea life is most abundant at a depth of about 30 feet (9.1 meters) of water.

It is farther off the shoreline, where the water is the most shallow.

STAGHORN CORAL

Fore Reef Zone

It is farthest away from the shoreline, where the water is deeper than in the other zones.

BOULDER CORAL

SAWTOOTH BARRACUDA

FIRE CORAL

LETTUCE CORAL

PLATE CORAL

DROP-OFF SLOPE

LEAFY CORAL

COLUMN CORAL

CACTUS CORAL

SEA FAN CORAL

How Coral Reefs Grow

At certain times of the year, after a full moon, many corals in a reef release eggs into the seawater. The eggs soon become floating baby corals, called planulae (PLAN•yuh•lie). When a planula (PLAN•yuh•luh) attaches itself to a reef or a hard surface, it forms into a coral polyp.

A tiny oval larva of a coral polyp is called a PLANULA.

It uses tiny appendages called cilia (SIL•ee•uh) to swim.

TOMATO ANEMONEFISH

CORAL EGG

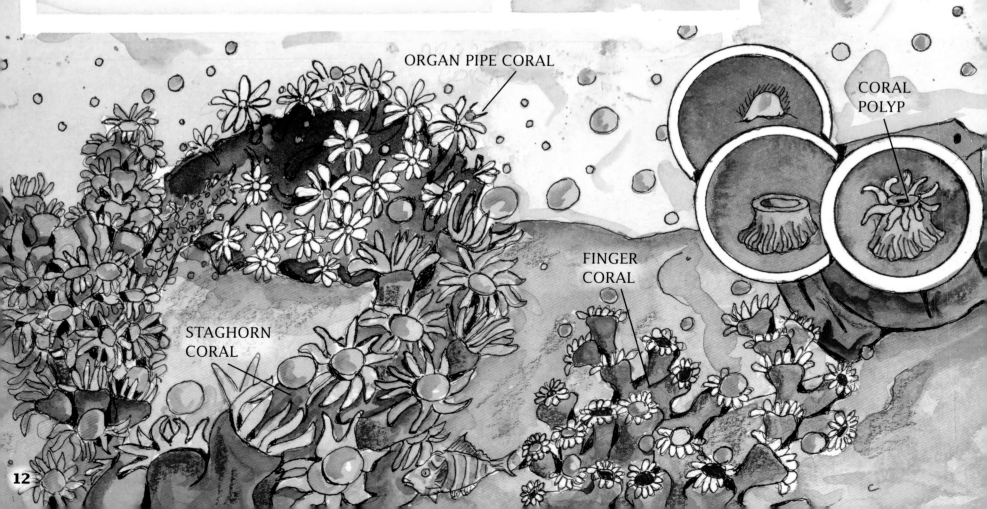

ORGAN PIPE CORAL

CORAL POLYP

FINGER CORAL

STAGHORN CORAL

SEA
SLUG

CHOCOLATE
SURGEONFISH

BROAD-BANDED
CARDINALFISH

SEA ROD CORAL

Over time each coral polyp reproduces another polyp connected to it. Every new polyp does the same thing. Gradually the coral structure becomes larger and larger.

LONG-SPINED
SEA URCHIN

ENCRUSTING CORAL

NEW CORAL POLYP

TUBE
SPONGE

Hard and Soft Corals

Hard Corals

The polyps of hard corals are held in place by a hard skeletal base. Above the base its body is soft with stinging tentacles surrounding its edge. To protect itself it will pull its soft body and tentacles inside its hard cup base. There are about 650 species of hard coral.

BANDED SEA SNAKE

BRANCHING CORAL

LEAFY CORAL

DOTTED BUTTERFLYFISH

HONEYCOMB CORAL

MUSHROOM CORAL

STAR CORAL

RUBY BRITTLE STAR

Polyp of Hard Coral

TENTACLES are used for protection and to catch prey.

MOUTH

SOFT BODY

HARD CUP BASE

Soft Corals

A soft coral attaches itself to a reef with its soft, sticky base. Soft corals have soft skeletons. They build beautiful structures by reproducing themselves. These structures are soft and move with the motion of the sea. There are about 1,800 known species of soft coral.

GREAT BARRACUDA

TREE CORAL

SULFUR DAMSELFISH

SPIRAL CORAL

BUSHY SOFT CORAL

Polyp of Soft Coral

TENTACLES

MOUTH

SOFT BODY

SOFT BASE

TIMOR SNAPPER

SEA FAN CORAL

SEA PEN CORAL

SEA WHIP CORAL

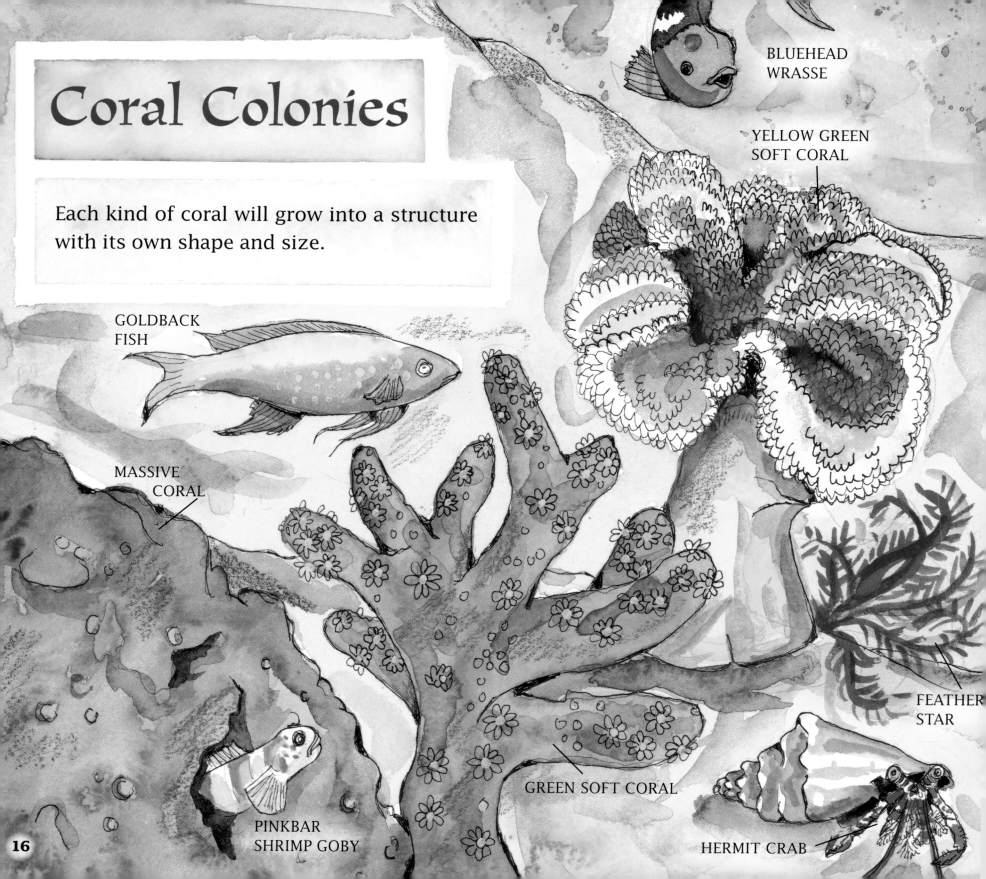

Coral Colonies

Each kind of coral will grow into a structure with its own shape and size.

BLUEHEAD WRASSE

YELLOW GREEN SOFT CORAL

GOLDBACK FISH

MASSIVE CORAL

FEATHER STAR

GREEN SOFT CORAL

PINKBAR SHRIMP GOBY

HERMIT CRAB

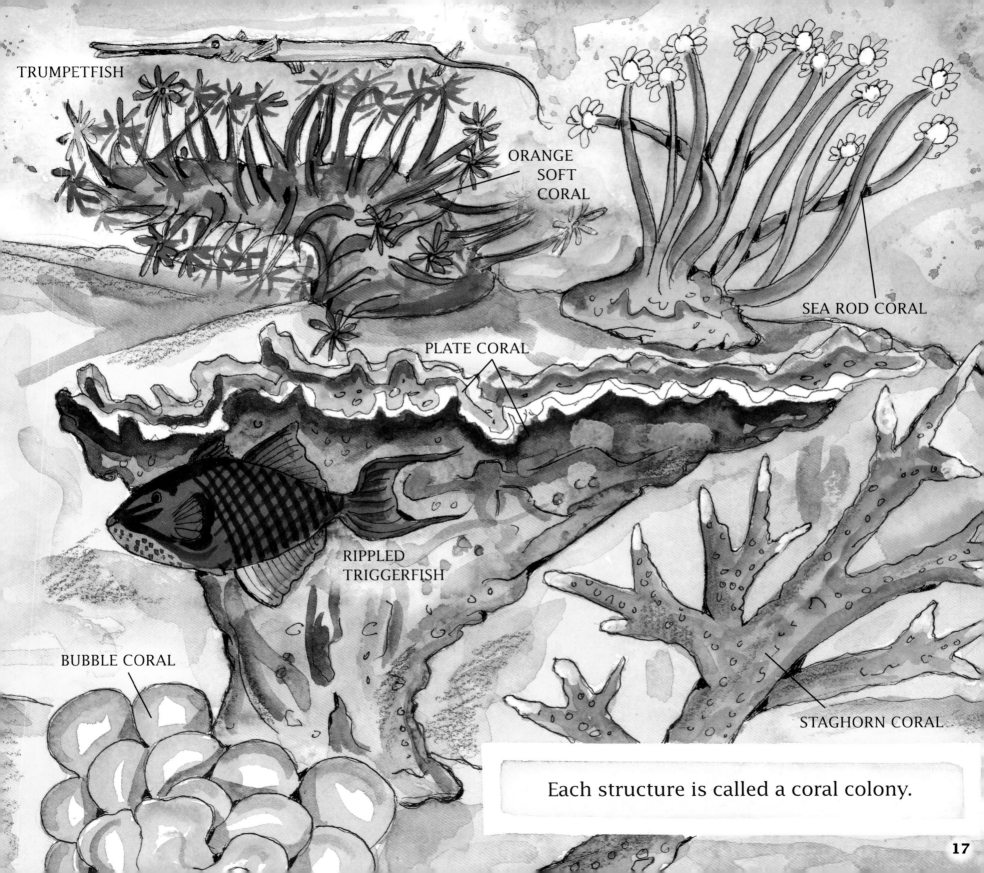

TRUMPETFISH

ORANGE
SOFT
CORAL

SEA ROD CORAL

PLATE CORAL

RIPPLED
TRIGGERFISH

BUBBLE CORAL

STAGHORN CORAL

Each structure is called a coral colony.

Daytime on a Coral Reef

Tiny plants called algae (AL•jee) live inside the polyps and make sugar. Sunlight makes this possible. This process is called photosynthesis (foe•toe•SIN•thuh•sus). Polyps feed on the sugar. The many different colors of algae create the many brilliant colors of a coral reef.

RED SOFT CORAL

BRAIN CORAL

FIRE CORAL

SPINY ROW CORAL

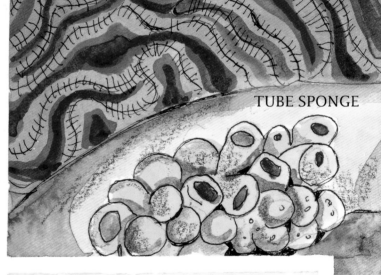

TUBE SPONGE

Photosynthesis in a Coral Reef

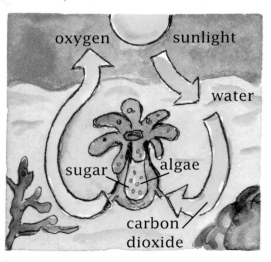

oxygen

sunlight

water

sugar

algae

carbon dioxide

Without photosynthesis there would be no oxygen, and all plant and animal life on Earth would cease to exist.

Polyp of Soft Coral

ALGAE

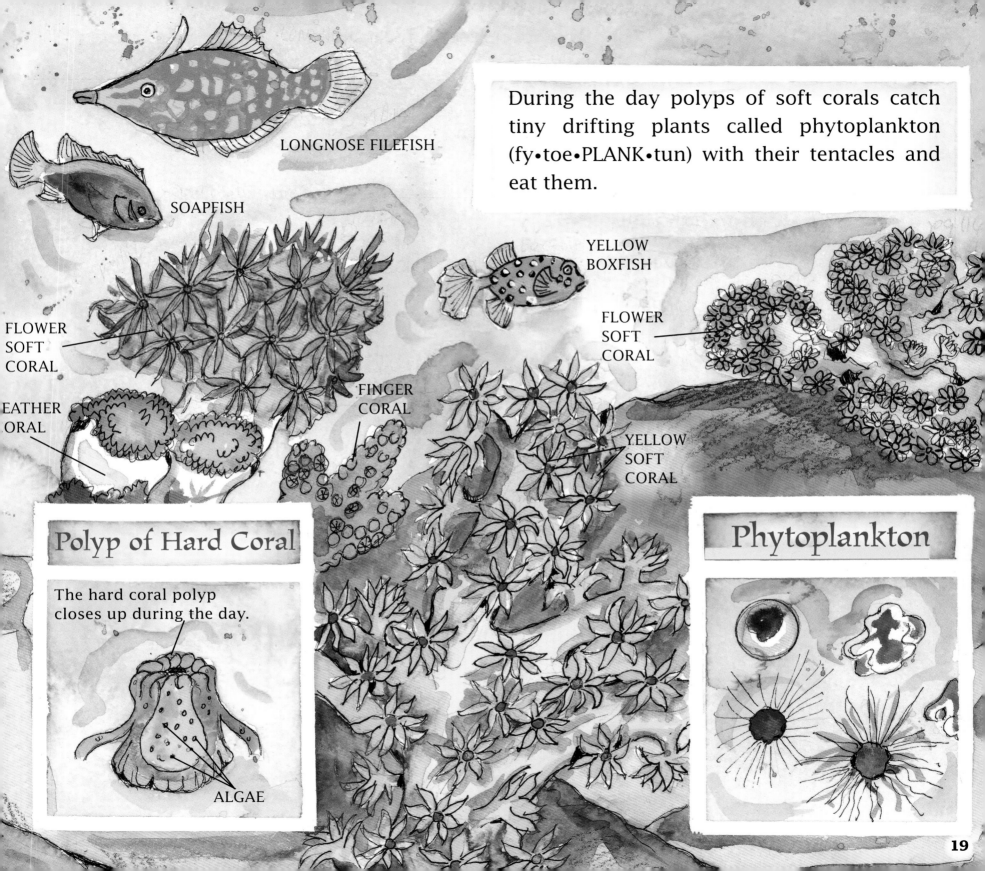

LONGNOSE FILEFISH

SOAPFISH

During the day polyps of soft corals catch tiny drifting plants called phytoplankton (fy•toe•PLANK•tun) with their tentacles and eat them.

YELLOW BOXFISH

FLOWER SOFT CORAL

FLOWER SOFT CORAL

EATHER ORAL

FINGER CORAL

YELLOW SOFT CORAL

Polyp of Hard Coral

The hard coral polyp closes up during the day.

ALGAE

Phytoplankton

Nighttime on a Coral Reef

The polyps of most hard corals eat only during the night. They catch extremely small animals called zooplankton (zoe•uh•PLANK•tun) with their tentacles and eat them. The polyps of soft corals continue to eat during the night.

BLUE-RINGED OCTOPUS

YELLOW SEA WHIP CORAL

ORANGE SOFT CORAL

ENCRUSTING CORAL

FINGER CORAL

Zooplankton

REEF WHITETIP SHARK

MASSIVE CORAL

CUP CORAL

Phytoplankton and zooplankton cannot be seen without a microscope.

CUTTLEFISH

MANTA RAY

FEATHER
STAR CORAL

BARRED
MORAY
EEL

LEATHER
CORAL

PINK
TREE
CORAL

COLUMN
CORAL

BUBBLE
CORAL

CORAL
CRAB

TABLE CORAL

CHRISTMAS
TREE
WORM

RED CAVE
CORAL

Polyp of Hard Coral

STAGHORN
CORAL

The reef is colorful during the day, but it is at its most colorful at nighttime, when the polyps of both hard and soft corals are open.

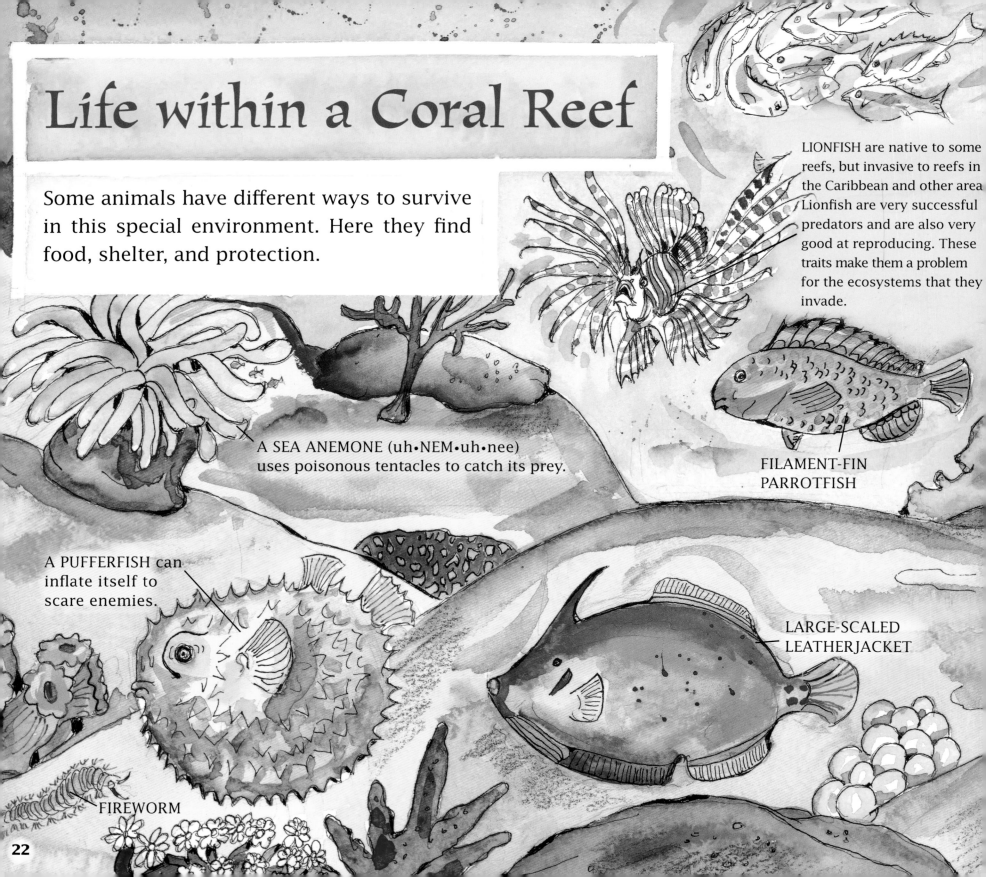

Life within a Coral Reef

Some animals have different ways to survive in this special environment. Here they find food, shelter, and protection.

LIONFISH are native to some reefs, but invasive to reefs in the Caribbean and other area Lionfish are very successful predators and are also very good at reproducing. These traits make them a problem for the ecosystems that they invade.

A SEA ANEMONE (uh•NEM•uh•nee) uses poisonous tentacles to catch its prey.

FILAMENT-FIN PARROTFISH

A PUFFERFISH can inflate itself to scare enemies.

LARGE-SCALED LEATHERJACKET

FIREWORM

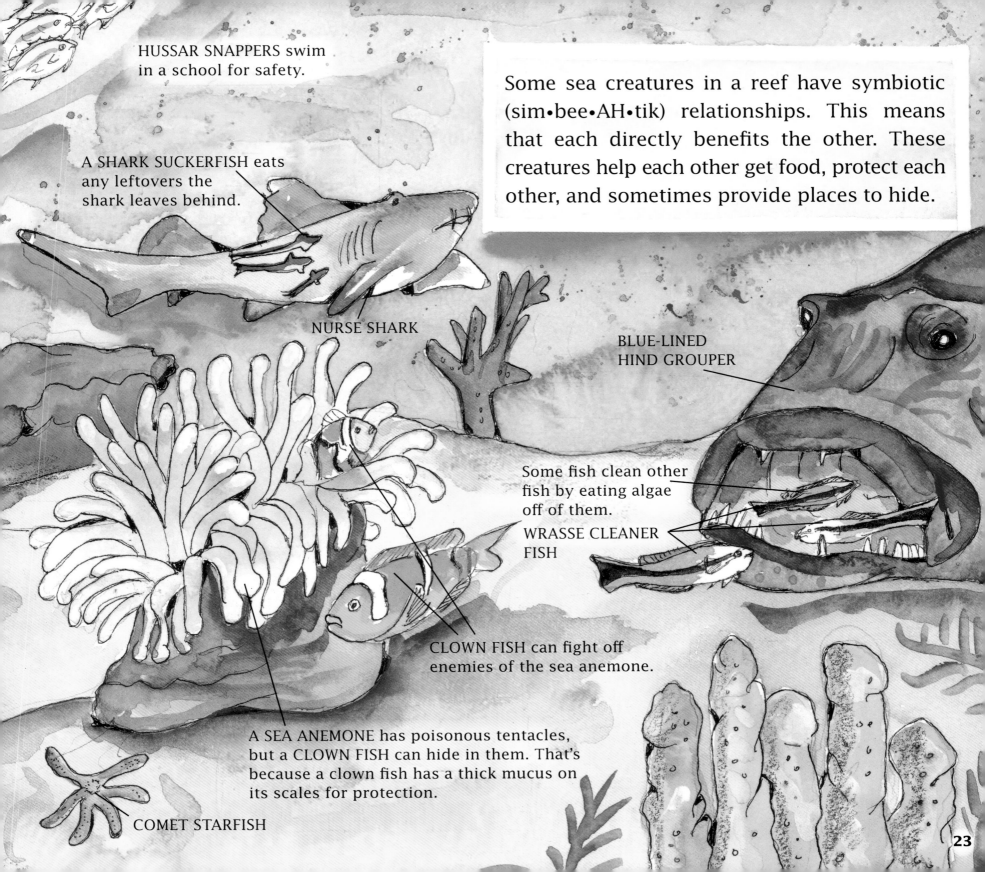

HUSSAR SNAPPERS swim in a school for safety.

A SHARK SUCKERFISH eats any leftovers the shark leaves behind.

Some sea creatures in a reef have symbiotic (sim•bee•AH•tik) relationships. This means that each directly benefits the other. These creatures help each other get food, protect each other, and sometimes provide places to hide.

NURSE SHARK

BLUE-LINED HIND GROUPER

Some fish clean other fish by eating algae off of them.

WRASSE CLEANER FISH

CLOWN FISH can fight off enemies of the sea anemone.

A SEA ANEMONE has poisonous tentacles, but a CLOWN FISH can hide in them. That's because a clown fish has a thick mucus on its scales for protection.

COMET STARFISH

23

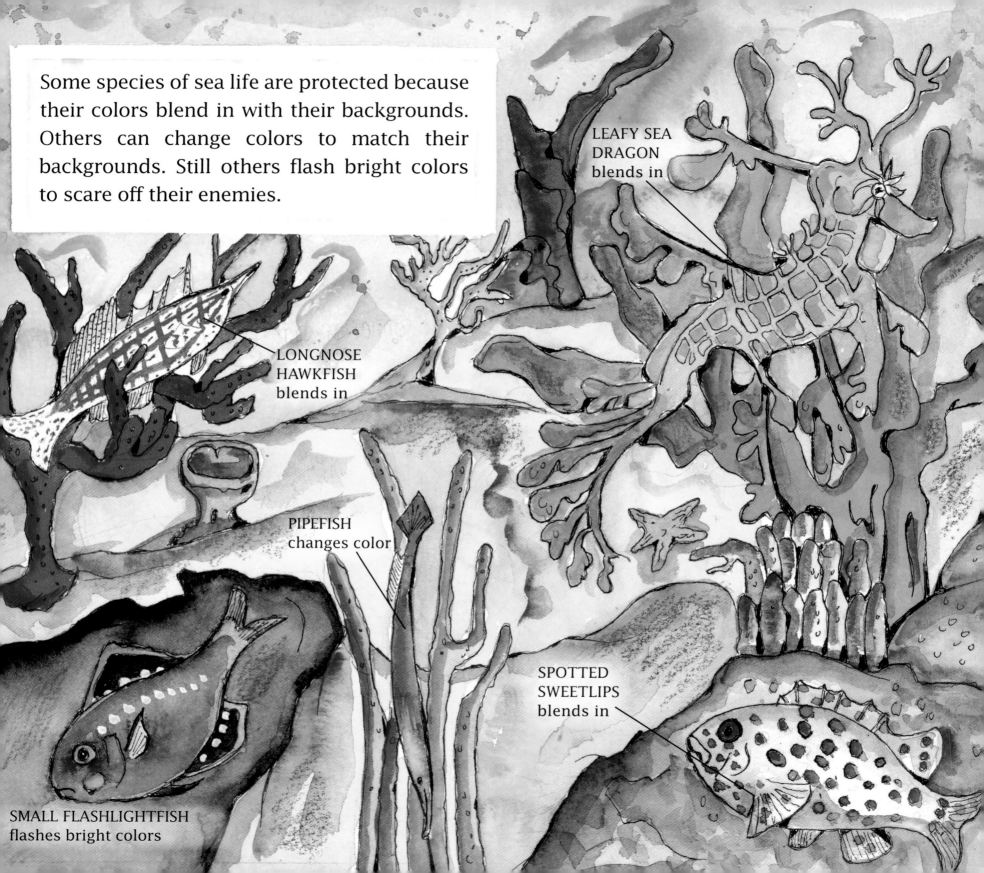

Some species of sea life are protected because their colors blend in with their backgrounds. Others can change colors to match their backgrounds. Still others flash bright colors to scare off their enemies.

LEAFY SEA DRAGON blends in

LONGNOSE HAWKFISH blends in

PIPEFISH changes color

SPOTTED SWEETLIPS blends in

SMALL FLASHLIGHTFISH flashes bright colors

Daytime

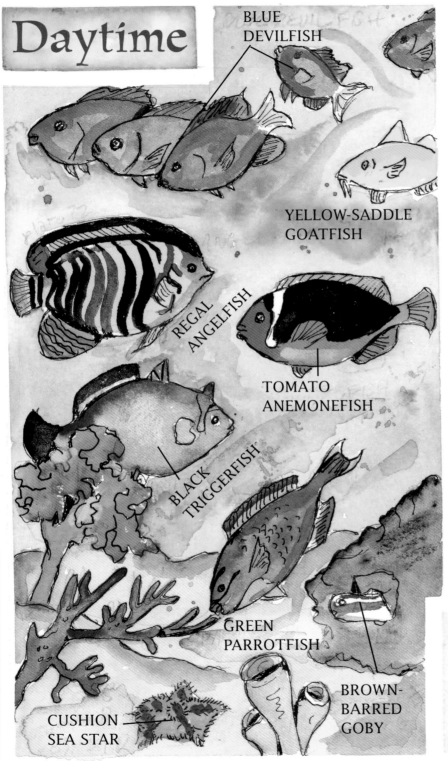

BLUE DEVILFISH

YELLOW-SADDLE GOATFISH

REGAL ANGELFISH

TOMATO ANEMONEFISH

BLACK TRIGGERFISH

GREEN PARROTFISH

BROWN-BARRED GOBY

CUSHION SEA STAR

Nighttime

GREEN SEA TURTLE

SANDBAR SHARK

MANTA RAY

GIANT BARRACUDA

BARRED MORAY EEL

A coral reef is a very busy place during the day. About two-thirds of all the reef creatures are active at this time. Others hide during the day and feed at night.

The Great Barrier Reef

The largest coral reef in the world lies off the east coast of Australia. The reef is about 1,430 miles (2,301 kilometers) long. The Great Barrier Reef is so large that astronauts can see it from outer space.

BENGAL SERGEANT DAMSELFISH

SEA SLUG

SPINY ROW CORAL

RACCOON BUTTERFLYFISH

MASSIVE CORAL

TUBE SPONGE

GIANT CLAM

CONE SNAIL

STAR CORAL

GREAT BARRIER REEF

AUSTRALIA

SOUTH PACIFIC OCEAN

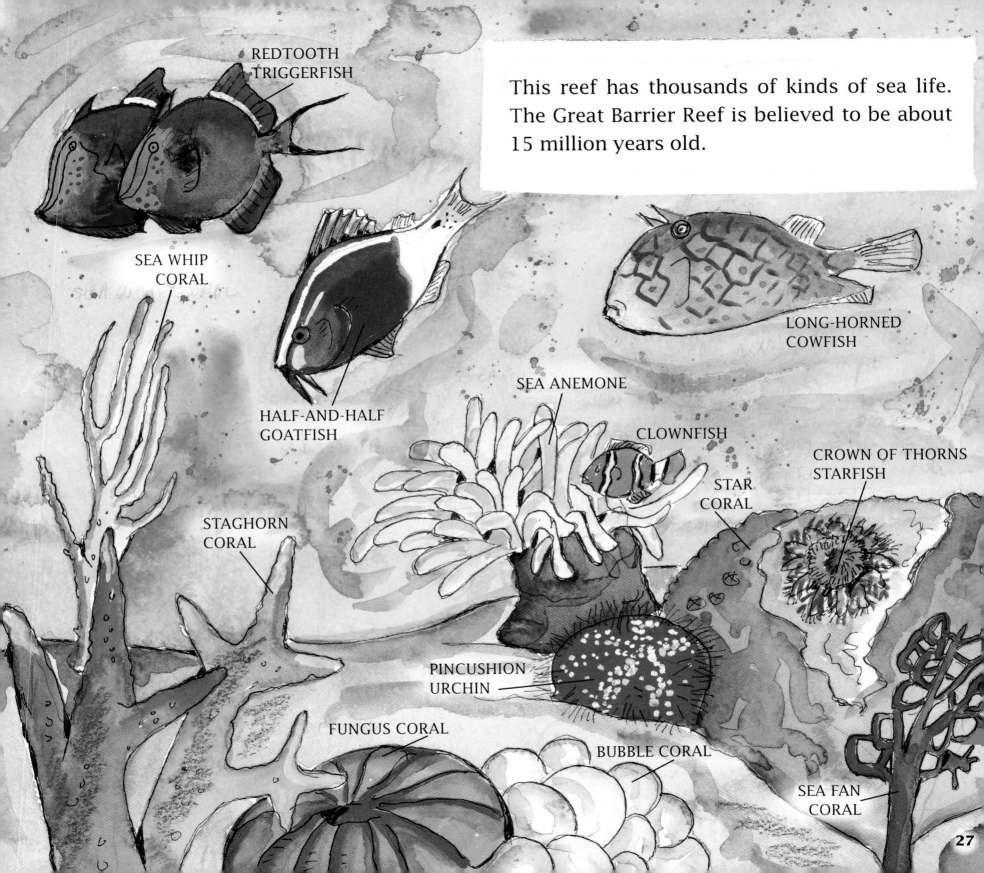

REDTOOTH TRIGGERFISH

This reef has thousands of kinds of sea life. The Great Barrier Reef is believed to be about 15 million years old.

SEA WHIP CORAL

LONG-HORNED COWFISH

SEA ANEMONE

HALF-AND-HALF GOATFISH

CLOWNFISH

CROWN OF THORNS STARFISH

STAR CORAL

STAGHORN CORAL

PINCUSHION URCHIN

FUNGUS CORAL

BUBBLE CORAL

SEA FAN CORAL

Look . . . but Don't Disturb!

Marine parks have been created to protect coral reefs. Some people use snorkels, and others use scuba-diving equipment, to explore the reefs.

SNORKEL

MASK

FLIPPERS

DO NOT TOUCH OR TAKE ANYTHING FROM THE REEF!

GLOBAL WARMING causes seawater temperatures to rise, damaging coral reefs and sealife.

Different colorful tiny plants called algae live in coral. This is what makes brilliant-colored coral reefs. Algae feeds coral by making sugar from sunlight. When sea temperatures rise, the coral becomes stressed and they release their colorful algae. This causes the coral to turn white and brittle. This process is called CORAL BLEACHING.

Coral reefs need to be protected. They actually help reduce global warming by using the carbon dioxide that is dissolved in the seawater and creating oxygen for all to breathe. Coral reefs everywhere are fragile and should not be disturbed.

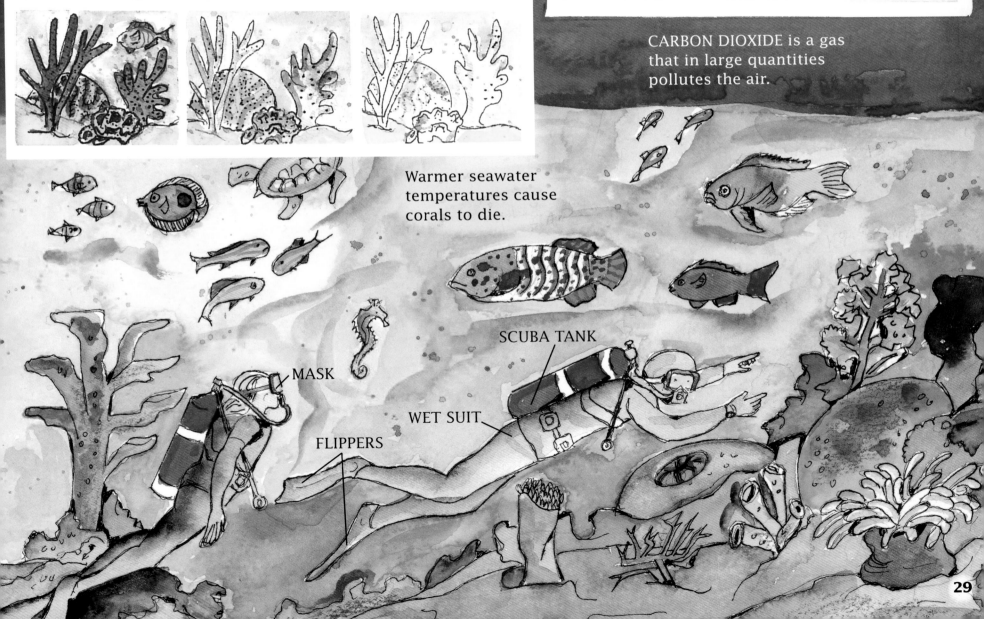

CARBON DIOXIDE is a gas that in large quantities pollutes the air.

Warmer seawater temperatures cause corals to die.

SCUBA TANK

MASK

WET SUIT

FLIPPERS

Many people like to visit sea aquariums to see coral reef exhibits.

EXHIBITS →

STING RAYS

Coral reefs are among the most beautiful and unusual places in the natural world. It is both fun and important to learn about them.

Coral Reefs

More sea creatures live in and around coral reefs than anywhere else in the world's oceans.

Coral reefs are hard. Throughout history many ships sank when they ran into coral reefs.

About 200,000 kinds of plants and animal life that live within coral reefs have been discovered and named.

The United States has only one percent of all known coral reefs.

Marine biologists believe there could be as many as 2,000,000 types of sea life inhabiting the world's coral reefs.

Australia, the Philippines, and Indonesia have about half of all known coral reefs.

Papahānaumokuākea Marine National Monument

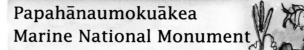

A giant clam can grow to be more than 4 feet (1.2 meters) wide and can weigh more than 500 pounds (226.8 kilograms).

In 2006 the United States designated the area around the Northwestern Hawaiian Islands a marine national monument. In 2007 it was renamed the Papahānaumokuākea (PAH-pah-HA-naoo-MOW-koo-a-KAY-uh) Marine National Monument. This monument includes 140,000 square miles (about 363,000 square kilometers) of ocean, islands, and reefs, an area larger than all U.S. national parks combined.

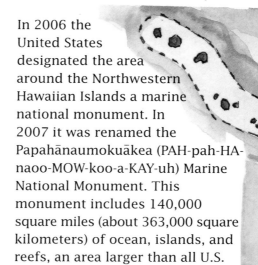

480 miles (773 kilometers) to HONOLULU

Coral bleaching has become much more common on reefs worldwide. In the 1980s, coral bleaching impacted reefs every 25 or 30 years. Today, reefs are impacted at least every 6 years. Some corals can recover from bleaching, but many do not.